W9-AUI-764

Jerome Lemelson

The Man Behind Industrial Robots

by Lucia Raatma

PEBBLE
a capstone imprint

Little Explorer is published by Pebble, an imprint of Capstone.
1710 Roe Crest Drive
North Mankato, Minnesota 56003
www.capstonepub.com

The name of the Smithsonian Institution and the sunburst logo are registered trademarks of the Smithsonian Institution. For more information, please visit www.si.edu.

Library of Congress Cataloging-in-Publication data is available on the Library of Congress website.
ISBN 978-1-9771-1412-9 (library binding)
ISBN 978-1-9771-1787-8 (paperback)
ISBN 978-1-9771-1416-7 (eBook PDF)
Summary: Gives facts about Jerome Lemelson, his life, and how his inventions changed the world.

Image Credits
Alamy: Gado Images, 14; AP Photo: Mark Elias, 20; Courtesy of the Lemelson Family: cover (top), 5, 6, 7 (left), 13, 15, 16, 19, 23 (bottom), 25 (top), 29; Getty Images: Archive Photos/Lawrence Thornton, 9, Gamma-Keystone/Keystone-France, 8; Shutterstock: Alpa Prod, 11 (right), asharkyu, cover (bottom), August Phunitiphat, 17 (bottom), Everett Historical, 7 (right), Jenson, 4, Mark Van Scyoc, 22, Pavel L Photo and Video, 21; Smithsonian Institution: Photo by Jeff Tinsley, 25 (bottom), 26; U.S. Army CCDC: 27; U.S. Patent and Trademark Office: 11 (left), 17 (top), 23 (top)

Design Elements by Shutterstock

Editorial Credits
Editor: Michelle Parkin; Designer: Sarah Bennett; Media Researcher: Svetlana Zhurkin; Production Specialist: Tori Abraham

Our very special thanks to Emma Grahn, Spark!Lab Manager, Lemelson Center for the Study of Invention and Innovation, National Museum of American History. Capstone would also like to thank Kealy Gordon, Product Development Manager, and the following at Smithsonian Enterprises: Ellen Nanney, Licensing Manager; Brigid Ferraro, Vice President, Education and Consumer Products; and Carol LeBlanc, Senior Vice President, Education and Consumer Products.

All internet sites appearing in the back matter were available and accurate when this book was sent to press.

Printed in the United States of America.
PA99

TABLE OF CONTENTS

Bold words are in the glossary.

INVENTING ROBOTS

Turn. Turn. Turn. This is the motion of a robotic arm. The arm is connected to a large machine in a car factory. The robot puts doors onto a long line of trucks.

Industrial robots put cars together on an assembly line.

Industrial robots help make products. They can paint, put parts together, or test finished products. An **inventor** named Jerome Lemelson helped make robots like these possible.

Jerome Lemelson

EARLY LIFE

Jerome Lemelson was born on July 18, 1923, in Staten Island, New York. His friends and family called him Jerry. As a young boy, Jerry was interested in airplanes and flying. He made **model** airplanes in his basement.

Jerome Lemelson (right) and his brother, Howard, made model planes together.

One of Jerome Lemelson's role models was inventor Thomas Edison.

After high school, Lemelson attended New York University. The United States entered World War II (1939–1945) in 1941. Lemelson left college to fight in the war. He joined the U.S. Army Air Corps.

The U.S. Army Air Corps is now called the U.S. Air Force.

After World War II ended, Lemelson returned to New York University. He studied engineering. This science taught him how airplanes and other machines worked.

A class at New York University in 1945

Outside the Library of New York University in the 1940s

Lemelson graduated from college in 1951. He earned a bachelor's degree and two master's degrees in engineering.

Lemelson worked for the Office of Naval Research as a student. He designed rocket and jet engines.

MACHINE VISION AND ROBOTS

Lemelson wanted to be an inventor. He was always thinking of ways to improve the things around him. One of his ideas was for a technology called machine vision. Machine vision used computers to analyze images from a video camera. Using this technology, a machine could "see" what it was doing.

Lemelson invented an early version of the cordless phone. He also invented a talking thermometer for the blind.

"I am always looking for problems to solve. I cannot look at a new technology without asking: How can it be improved?"

—Jerome Lemelson

Lemelson applied for a **patent** for machine vision on Christmas Eve in 1954. His patent was approved more than 30 years later! His idea eventually led to the **bar code** reader. Today, bar code readers are used in stores all over the world.

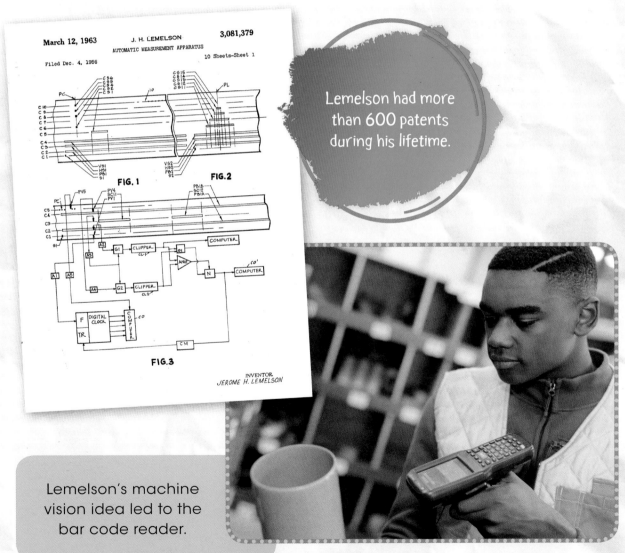

Lemelson had more than 600 patents during his lifetime.

Lemelson's machine vision idea led to the bar code reader.

Lemelson married Dorothy "Dolly" Ginsburg in 1954. They had two sons—Eric and Robert. They lived in Metuchen, New Jersey. Lemelson worked many hours a day on his inventions. He did not mind the long hours. But he did not like working for other people or companies.

Lemelson wanted to work for himself. He decided to work full time as an inventor. Lemelson left his job as an **engineer**. He worked on his inventions at home.

Jerome and Dolly Lemelson in 1954

Lemelson with his sons Eric (left) and Robert (right) in 1969

The Lemelsons lived near Menlo Park. This is where Thomas Edison came up with some of his most famous inventions.

Workers stood outside Edison's lamp factory at Menlo Park in 1880.

An important part of being an inventor is **licensing** one's ideas. People can't buy new items if they do not know they exist. Lemelson wanted to promote his inventions himself. He founded the Licensing Management Corporation to sell his ideas.

During this time, Dolly ran her own interior design business. She supported the family so Lemelson could work on his inventions.

Lemelson with an industrial robot in 1994

Lemelson's most successful invention was the universal robot. The robot would use machine vision to study a task. Then the robot could figure out the best way to complete it.

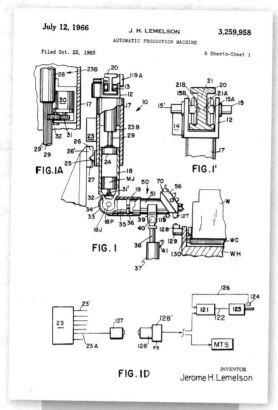

July 12, 1966 J. H. LEMELSON 3,259,958
AUTOMATIC PRODUCTION MACHINE
Filed Oct. 22, 1965 6 Sheets-Sheet 1

FIG.IA

FIG.I'

FIG.I

FIG.ID

INVENTOR.
Jerome H. Lemelson

One of Lemelson's patents. It led to universal robots.

Universal robots could pick up, measure, and weld machine parts. They even made sure all the parts were put together correctly. The universal robot helped make car **assembly lines** faster and more efficient. This also led to a warehouse system that helped companies keep better track of their goods. The warehouse system is still used by many industries today.

Universal robots are still used in factories today.

TAKING NOTES

Lemelson came up with ideas for new inventions all the time. He kept a notebook with him. He even took the notebook to bed in case he dreamed up a great invention.

When Lemelson wrote down an idea in his notebook, he would ask a person standing nearby to sign the page. Lemelson used these signed pages as proof that he came up with his own ideas. He did not want people to think he took someone else's inventions.

"Don't cut back on basic research. It's necessary to the future of this country."

—Jerome Lemelson

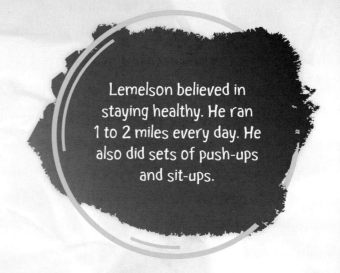

Lemelson believed in staying healthy. He ran 1 to 2 miles every day. He also did sets of push-ups and sit-ups.

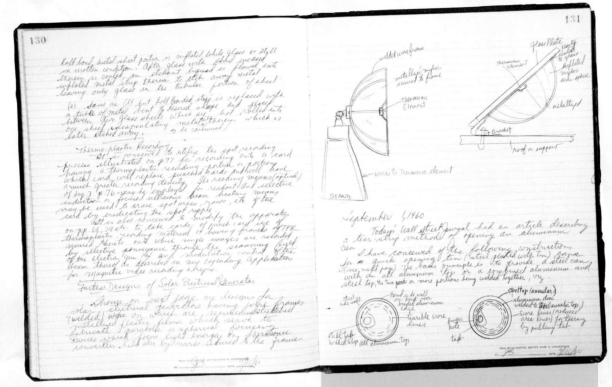

Pages from one of Lemelson's invention notebooks

Lemelson spent hours each day working on his new inventions. He would sketch out his ideas. He would alter the details.

Lemelson wrote all his ideas in notebooks he carried with him.

Lemelson also invented toys. One fun invention was a kind of roller skate. Eventually this idea became in-line skates. Another invention was a looped track for race cars.

Lemelson invented a type of track for toy race cars.

20

In 1977, Lemelson applied for a patent for a camcorder. It was rejected. People thought the technology was impossible.

Lemelson invented a type of cassette recorder. The company Sony bought the patent for his invention in 1974. They used Lemelson's idea to create the Walkman. Eventually, this led to MP3 players.

Lemelson filed patents to protect his inventions. This way, no one else could take credit for them and make money. But patents can take a long time to get approved. Over the years, Lemelson learned that other companies were using his ideas. He filed **lawsuits** to protect his inventions.

UNITED STATES PATENT AND TRADEMARK OFFICE

Patents are reviewed by the United States Patent and Trademark Office in Arlington, Virginia.

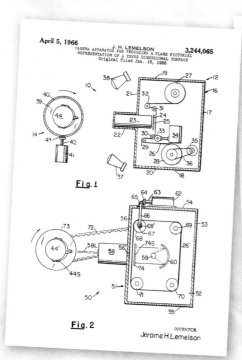

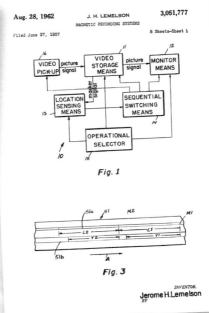

One of Lemelson's lawsuits went on for more than 22 years!

Lemelson's ideas led to many inventions, including electronics and toys.

GIVING BACK

For most of his life, Lemelson was not a wealthy man. He worked hard, but applying for patents was expensive. A fee is charged each time a person applies for a patent. In his later years, Lemelson received large payments from some companies that had used his patented ideas. He used his money to help other inventors succeed.

In 1993, Jerome and Dorothy Lemelson created the Lemelson Foundation. This group offers programs to help young people become inventors and business owners. It encourages them to come up with new ideas. It also teaches people how to sell their inventions.

Lemelson and his wife created the Lemelson Foundation.

In the 1990s, Lemelson worked with students interested in creating their own inventions.

The Jerome and Dorothy Lemelson Center for the Study of Invention and Innovation opened in 1995.

In 1995, the Lemelson family created the Jerome and Dorothy Lemelson Center for the Study of Invention and Innovation. It is located at the Smithsonian Institute in Washington, D.C. This center helps people solve problems and improve their communities through invention. The center shows the history of invention and promotes new ways of thinking about inventions.

Students looking at inventions on display at the Lemelson Center

In the mid-1980s, Jerome and Dorothy Lemelson moved to Princeton, New Jersey. Ten years later, they settled in Lake Tahoe, Nevada. They enjoyed the fresh air and being outdoors.

In 1996, Lemelson learned that he had liver cancer. He wanted to do what he could with the time he had left. He spent even more time inventing. He filed almost 40 patent applications in his last year. Jerome Lemelson died on October 1, 1997. Today, not many people have heard of Jerome Lemelson. But we use versions of his inventions every day.

Inventor Paul McCready (left) and activist Ralph Nader (center) met with Lemelson in 1995.

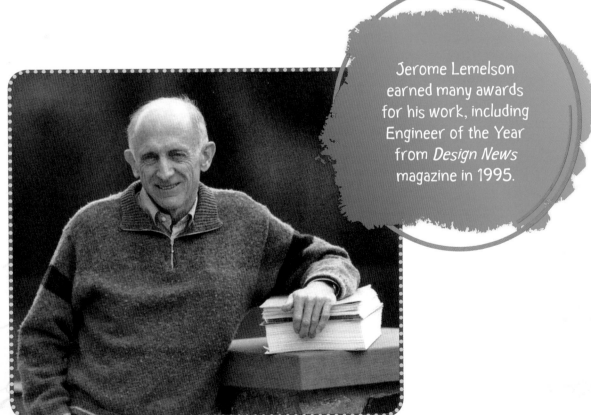

Jerome Lemelson earned many awards for his work, including Engineer of the Year from *Design News* magazine in 1995.

GLOSSARY

assembly line (uh-SEM-blee LYN)—moving belts bring work from one person to the next until the product is made; cars and car parts move along an assembly line

bar code (BAR KODE)—a series of bars or stripes that give coded information

engineer (en-juh-NEER)—someone trained to design and build machines, vehicles, bridges, roads, or other structures

inventor (in-VENT-ohr)—someone who thinks up and creates something new

lawsuit (LAW-soot)—a legal action or case brought against a person or group in a court of law

license (LYE-suhnss)—a document that gives official permission to do something

model (MOD-uhl)—something that is made to look like a person, an animal, or an object

patent (PAT-uhnt)—a legal document giving the inventor sole rights to make and sell an item he or she has invented

CRITICAL THINKING QUESTIONS

1. Why do inventors need patents for their ideas?

2. Why do you think new inventions are important?

3. Imagine you are an inventor. What would you create? Why?

READ MORE

Clay, Kathryn. *Robots on the Job.* North Mankato, MN: Capstone Press, 2015.

Lepora, Nathan. *Robots.* New York: DK Publishing, 2018.

Waxman, Laura Hamilton. *Cool Kid Inventions.* Minneapolis: Lerner Publications Company, 2020.

INTERNET SITES

Lemelson Center
https://invention.si.edu/

The Lemelson Foundation
https://www.lemelson.org/

INDEX